How Plants Make Food

BY MARTIN J. GUTNIK

ILLUSTRATED BY SAM SHIROMANI

CHILDRENS PRESS, CHICAGO

Library of Congress Cataloging in Publication Data

Gutnik, Martin J
How plants make food.

SUMMARY: Uses experiments to explain how plants use carbon dioxide and water in conjunction with light to make food.
1. Photosynthesis—Experiments—Juvenile literature.
[1. Photosynthesis—Experiments] I. Shiromani, Sam.
II. Title.
QK882.G9 581.1′ 3342 75-37602
ISBN 0-516-00526-X

Printed in the United States of America

1 2 3 4 5 6 7 8 9 10 11 12 R 78 77 76

To my cousins,
Brenda and Marvin Kass

CONTENTS

FOREWORD

Photosynthesis is the process by which green plants make food. It comes from two words: *photo*, which means light, and *synthesis*, which means putting together.

Green plants put things together to make food. They use light to help them. Green plants are the only living things on earth that can make their own food.

All animals and human beings need green plants for food. Without green plants there could be no food.

Green plants also make *oxygen*. Oxygen is the gas we breathe. The plants make oxygen at the same time they make food. If green plants did not make oxygen, we could not live. We would not have enough oxygen to breathe.

This book will help you to learn about the raw materials a plant uses to make food. Raw materials are things used to make something. (Wood is one of the raw materials used to make chairs.)

A plant uses two raw materials. It uses carbon dioxide (CO_2), which is a gas found in our air. It also uses water. The plant uses carbon dioxide and water to make food. The plant's food is sugar.

A green plant also needs energy. Energy is the power to do work. Gas is energy for a car. The plant uses light for its energy. Light helps the plant make food.

You are going to learn more about the raw materials and energy that plants use to make food.

DO PLANTS REALLY NEED LIGHT, WATER, AND CARBON DIOXIDE?

Brenda wanted to know how plants make food. Her teacher had told her that plants make their own food. Brenda wanted to learn how they do this.

Brenda saw that when her mother baked a cake she started with flour, sugar, water, and eggs. Her mother mixed these together. After mixing, Brenda's mother put the batter in the oven to bake. From these steps a cake would be made.

Brenda learned that a plant makes food the way her mother makes a cake.

Brenda's teacher told her that a plant starts with water and carbon dioxide gas. Carbon dioxide (CO_2) is a gas found in our air. The plant uses light as its oven. From these things the plant makes sugar and oxygen. Oxygen is a gas we breathe.

Brenda wanted to do an experiment. She wanted to prove that green plants need water, carbon dioxide, and light to make food. She would use four plants in her experiment. One plant would get no water. Another plant would get no light. The third plant would get no carbon dioxide gas. The fourth plant would get everything.

Things Brenda needed:

1. Four small flower pots.
2. Four small green plants (zebrina, geranium, or coleus).
3. Soda lime.
4. Water.
5. A large-mouthed jar or bell jar (large enough to fit over one of the potted plants).
6. Soil.
7. A scoop.
8. A spoon.
9. Masking tape.
10. A pencil.
11. Paper.

What Brenda Did

Brenda put four small flower pots in front of her. She tore off four pieces of masking tape. She put one piece of masking tape in front of each pot.

On the first piece of tape she wrote: POT 1—NO WATER. This plant would get no water. Brenda put that piece of tape on the first pot.

On the second piece of tape she wrote: POT 2—NO LIGHT. This plant would get no light.

On the third piece of tape Brenda wrote: POT 3—NO CO_2 (carbon dioxide). This plant would get no carbon dioxide.

On the fourth piece of tape Brenda wrote: POT 4—CONTROL. This plant would get everything it needed. It would get water, light, and carbon dioxide.

Brenda now put soil in each pot. She was very careful to put the same kind of soil in each pot. She also put the same amount of soil in each pot.

Brenda put a plant in each pot. She planted each plant very carefully. She dug a hole with a spoon. She put the plant in the hole and covered it. She made sure that all the roots were covered. This is very important. If the roots are not covered, the plants will die.

After all the plants were potted, Brenda put each of them in its proper place.

POT 1—NO WATER: Brenda put this pot where it would get light. She did not water it. This plant would never be watered.

POT 2—NO LIGHT: Brenda watered this pot well. She put it in a dark closet so it would get no light. This pot was to be watered whenever the soil got dry.

NO CO2
SODA LIME

POT 3—NO CARBON DIOXIDE: This plant was put where it would get light. Brenda had to do some special things with this pot. She filled a little cup with soda lime. Soda lime sucks in carbon dioxide gas. It would stop plant 3 from getting carbon dioxide. Brenda put the little cup next to POT 3. She watered POT 3. It was watered very well. Now POT 3 was covered with a large-mouthed jar. This would keep the fresh air out. POT 3 did not have to be watered every day. The jar would keep in the water that Brenda had put in before she covered the pot.

POT 4—CONTROL: Brenda put POT 4 where it would get light. She watered it well. This pot would be watered whenever the soil was dry.

Brenda took care of the plants for about two weeks. Then she put all the plants on a table. She wanted to see what had happened.

What Did Brenda See?

Brenda looked at all four plants. She made a chart. On the chart she wrote down what each plant looked like.

POT 1 NO WATER	POT 2 NO LIGHT	POT 3 NO CARBON DIOXIDE	POT 4 CONTROL
The plant in this pot was very wilted. Its color was bad. The plant was very sick.	The plant in this pot had hardly any color. It was sick.	The plant in this pot was wilted. The color was not healthy. It was sick.	The plant in this pot was very green. Its leaves were firm. Its stem was firm. It was very healthy.

From her chart, Brenda saw that the plants in POTS 1, 2, and 3 were sick. The only plant that was healthy was the plant in POT 4. This was because POT 4 got water, light, and carbon dioxide

Brenda had learned that if a plant does not get water, light, and carbon dioxide, it cannot make food. It cannot be healthy.

PLANTS AND CARBON DIOXIDE (CO_2)

Brenda wanted to learn more about plants and carbon dioxide. She could not believe that plants really use this gas to make food.

Brenda's teacher told her that we breathe out carbon dioxide. We breathe in fresh oxygen. The plant makes food for us to eat. The plant makes oxygen for us to breathe. We give the plant carbon dioxide to make food and oxygen.

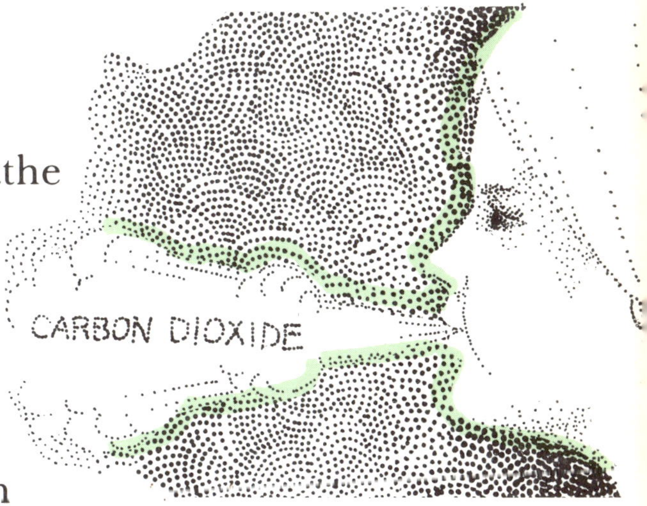

Brenda's teacher told her about a chemical called bromothymol blue. It turns yellow when it is mixed with carbon dioxide gas. If the carbon dioxide is taken away, it turns blue again.

Brenda thought that if a plant uses carbon dioxide, then it could make yellow bromothymol blue turn blue again. She did an experiment to find out.

Things Brenda needed:

1. Bromothymol blue solution. This is a mixture of bromothymol blue and water. You may get this from a drug store, high school science department, or a chemical supply house.
2. A straw.
3. Four test tubes with corks that fit on top. Clay may be used in place of corks.
4. Three elodea plants. These are water plants that can be found at most pet or fish stores.
5. A glass jar.
6. A funnel.
7. A test tube rack.

What Brenda Did

Brenda poured some bromothymol blue solution into a glass jar. She made sure that she put enough solution in the jar to fill four test tubes.

Brenda put a straw in the bromothymol blue solution and blew gently. She blew into the straw. This made bubbles in the solution. Brenda blew until the solution turned from blue to yellow.

When the solution was yellow, Brenda was ready to pour it into the test tubes.

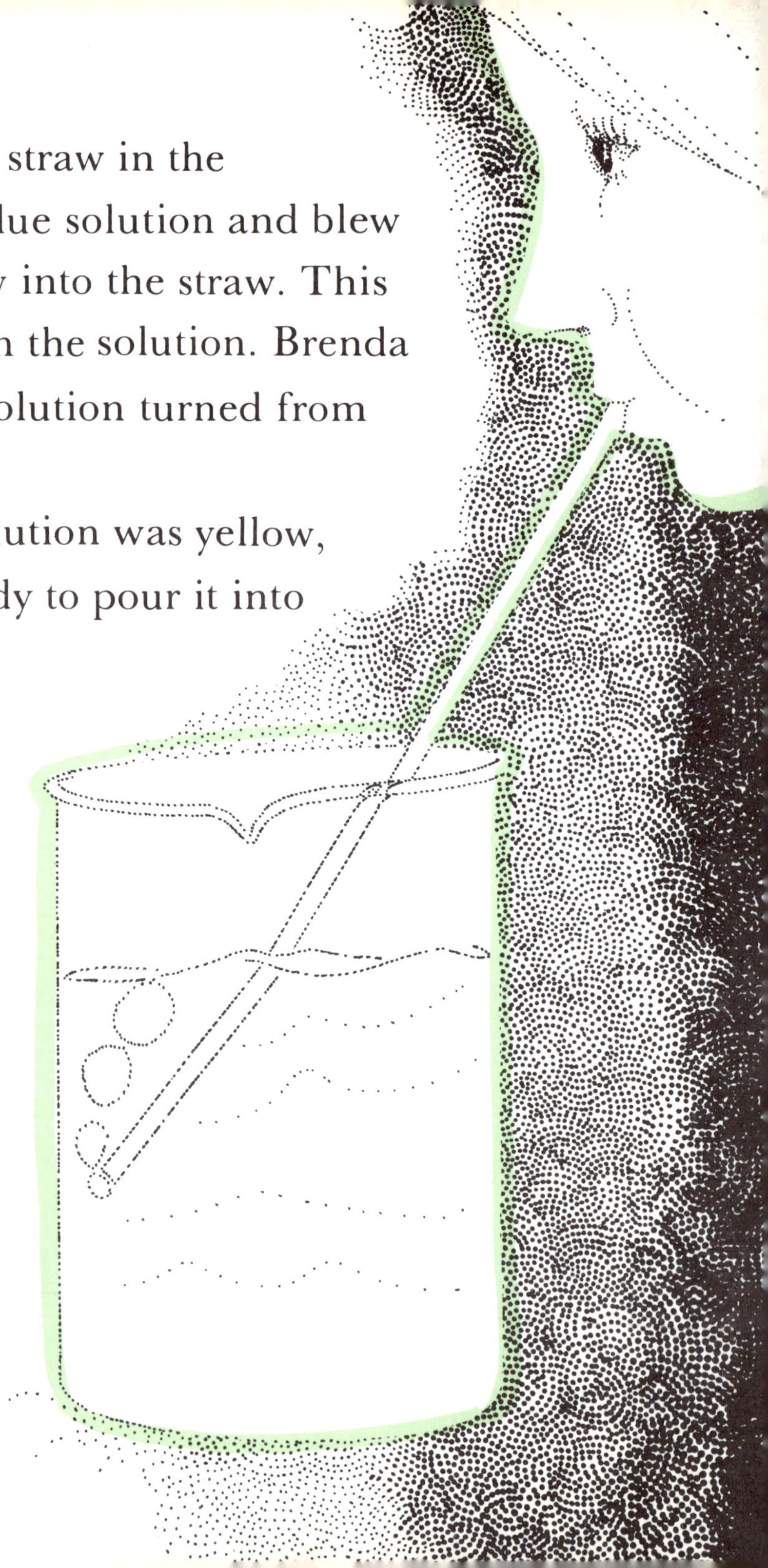

She used her funnel to fill one test tube at a time.

Brenda put an elodea plant in each of three test tubes. After she put in the plants, she corked the test tubes. She did this so no air could get in. When a test tube was corked she put it back into the test tube rack.

She did not put a plant into the fourth test tube. When it was filled, Brenda just put in a cork and put it in the rack.

Brenda put all four tubes in a place where they would get a lot of light. She watched them for ten days.

What Did Brenda See?

Brenda hoped that this experiment would prove that plants use carbon dioxide.

The three test tubes with the elodea plants turned blue. The tube that did not have a plant stayed yellow.

Brenda's teacher told her that this proved that plants use carbon dioxide gas. The test tubes with the plants were blue. They were blue because the plants had used the carbon dioxide in the water. When all the carbon dioxide in the water was gone, the solution changed from yellow to blue.

The test tube without a plant stayed yellow. This was because it did not have a plant to use up the carbon dioxide.

BLUE
BLUE
BLUE
YELLOW

PLANTS AND WATER: TRANSPIRATION

Brenda had learned from her first experiment that plants need water. She discovered that plants take in water through their roots. She wondered what happened to the water after the plant used it. She wondered where it went.

Brenda's teacher told her that plants lose water. They lose it all the time. He said that is called *transpiration.* This is when plants lose water from their leaves.

Brenda decided to do an experiment to see if plants usually lose water from their leaves.

Things Brenda needed:

1. One potted green plant (coleus, geranium, or zebrina).
2. One pot of soil—no plant.
3. Two large-mouthed jars large enough to fit over the potted plants.
4. Aluminum foil—enough to cover both pots and soil completely.
5. Masking tape.
6. Watering pitcher.
7. Pencil.

What Brenda Did

Brenda watered the pot of soil very well. This was the pot that had no plant in it. She watered it until it would hold no more water.

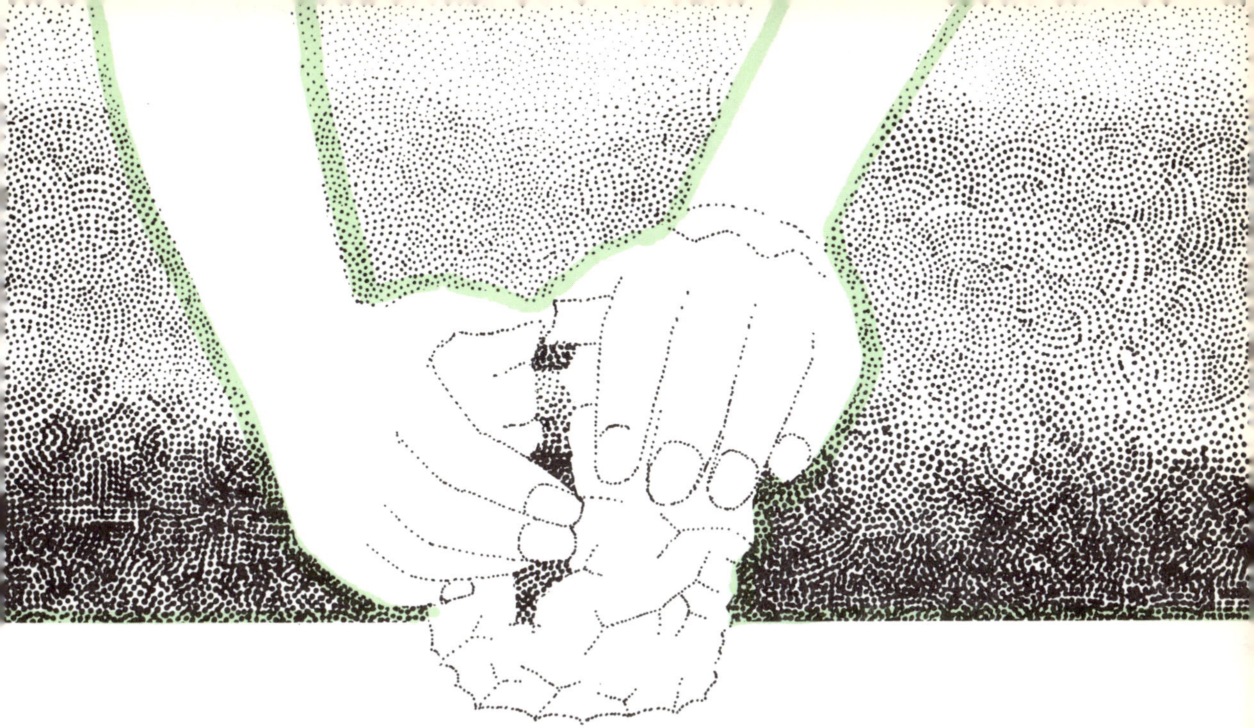

Then she covered thc pot with aluminum foil. She covered the whole pot. She covered the soil on top. The whole pot was covered with foil.

Brenda was very careful not to tear the foil. It had to be watertight to make sure that no water could get out.

Brenda now put the jar over the pot. She put masking tape on the jar. She wrote "JAR A" on the masking tape. The jar was put in a place where there was plenty of light.

Now Brenda watered the pot with the green plant. She watered it very well. She covered the pot and soil with aluminum foil. She wrapped the foil around the stem of the plant. She made sure that the pot and soil were watertight. If any water got out it would have to go through the plant.

Brenda put the other jar over this pot. She put masking tape on this jar. She labeled the jar "JAR B." Brenda put JAR B on the window sill next to JAR A. She let the pots stand for three days. She did not lift the jars during the three days.

What Did Brenda See?

When Brenda looked at the pots she knew that plants do lose water.

JAR A (the pot with no plant) was dry. JAR B (the pot with the plant) had drops of water all over it. There was only one place the water could have come from—the plant. Brenda had proved that plants do lose water from their leaves.

WORDS YOU SHOULD KNOW

Bromothymol blue solution (BROH moh thye mahl bloo suh LOO shun) A chemical used in testing whether a plant uses carbon dioxide gas. It turns yellow when it is mixed with carbon dioxide gas.

Carbon dioxide gas (CO_2) (KAR bun die AHK side gas) A gas found in the air. Plants use it as a raw material to make food.

Chlorophyll (KLOHR oh fill) A green coloring in plants. It absorbs (sucks in) the sun's energy and uses it to help the plant make food. Chlorophyll is the only thing on earth that can absorb the sun's energy to make food.

Chloroplast (KLORH oh plast) A plant tissue that holds chlorophyll.

Energy (EHN uhr gee) The power to do work.

Light (lite) The energy plants use to make food.

Phloem (FLOW ehm) Tubes (veins) in a plant that carry the sugar that the plant makes.

Photosynthesis (foh toh SIN thuh sihs) Made up of two words: *photo,* which means light, and *synthesis,* which means putting together. A plant puts water and carbon dioxide together. It uses light as its helper. It also uses chlorophyll as a helper. When these things are put together they make sugar and oxygen.

Raw material (rahw mah TEER ee ahl) A substance in its natural state.

Root Part of the plant that grows down into the ground. It holds the plant in the ground. It also gives the plant all its water and food.

Root hair It is like a hair. It comes out of the root like a branch. It absorbs (sucks in) the water and food for the plant.

Stele (STEE leh) The center of a root. It holds the veins that carry water and sugar.

Soda lime (SOH dah lime) A chemical that absorbs (sucks in) the carbon dioxide gas that is in the air around it.

Solution (suh LOO shun) A substance that is mixed with water.

Support tissue (suh POHRT tish yew) These cells are like your skeleton. They are long and narrow. They give the plant its form and shape. These cells also contain chloroplasts.

Tissue (TISH yew) A group of cells that look the same and do a certain job.

Transpiration (TRAN spih ray shun) Occurs when plants lose water from their leaves.

Water (H_2O) (WAH ter) A liquid that plants use as a raw material to make food.

Xylem (ZYE lehm) Tubes (veins) that carry water and food to the plant.

ACKNOWLEDGMENTS

Barbara Berg—Third grade student, Atwater School, Shorewood, Wisconsin

Margaret (Mrs. Bruce) Berg—Illustrator, science aide, Atwater School, Shorewood, Wisconsin

John Tranetski—Graduate student, University of Wisconsin, Milwaukee

Forest Stearns—Professor of Ecology, University of Wisconsin, Milwaukee

Timothy Stearns—Sixth grade student, Atwater School, Shorewood, Wisconsin

About the Author:

Martin Gutnik, an innovative elementary school science teacher, lives with his wife and two young children in Milwaukee, Wisconsin. His three latest books for Childrens Press, *How Plants Are Made, How Plants Make Food,* and *What Plants Produce,* are based on a series of experiments on photosynthesis that he created for his students to perform in the classroom. His first three *First Experiments in Science & Nature* books were based on the same type of experiments in the field of ecology and pollution. Through Mr. Gutnik spends much of his time helping his students learn to build ecosystems, develop film, and dissect frogs, his life is not totally surrounded by science. He also collects records from the fifties and enjoys singing folk songs as he accompanies himself on the guitar, which he taught himself to play.

About the Artist:

Born in India, Sam Shiromani has made Chicago his home. A dropout from the world of advertising, he devotes most of his time to free-lancing art and photography.

DATE DUE
